# SpringerBriefs in Computer Science

*Series Editors*
Stan Zdonik
Peng Ning
Shashi Shekhar
Jonathan Katz
Xindong Wu
Lakhmi C. Jain
David Padua
Xuemin Shen
Borko Furht
V.S. Subrahmanian
Martial Hebert
Katsushi Ikeuchi
Bruno Siciliano

T0211994

For further volumes:
http://www.springer.com/series/10028

Qing Zhou • Long Gao • Ruifang Liu
Shuguang Cui

# Network Robustness under Large-Scale Attacks

Springer

Qing Zhou
Texas A&M University
College Station, TX, USA

Long Gao
Hitachi Research Lab
Santa Clara, CA, USA

Ruifang Liu
Beijing University of Posts
    and Telecommunications
Beijing, P. R. China

Shuguang Cui
Texas A&M University
College Station, TX, USA

ISSN 2191-5768          ISSN 2191-5776 (electronic)
ISBN 978-1-4614-4859-4     ISBN 978-1-4614-4860-0 (eBook)
DOI 10.1007/978-1-4614-4860-0
Springer New York Heidelberg Dordrecht London

Library of Congress Control Number: 2012947522

Printed on acid-free paper

Springer is part of Springer Science+Business Media (www.springer.com)

# Preface

The robustness of a communication network under attacks is of prominent importance to both civilian service providers and military system operators. In this monograph, we start with literature overview of the network reliability study for networks under various attacks and then focus on the robustness of networks under large-scale physical attacks. In particular, we study the area-attack case, where each attack kills all the nodes and links that are touched by the attack area. Such a scenario can be a result of attacks from large-scale power outages or weapons of massive destruction, and, in general, it belongs to the category of networks under correlated attacks. Specifically, for the network under consideration, we assume that the nodes are deployed over a unit area according to a Poisson point process, and consider both the traditional random network model where each node pair is connected with a certain probability and the range-limited random network model where a node pair is connected with a certain probability only if the two nodes are within a certain range. The attack area is modeled as a small dish with radius $r$ and randomly located within the unit area. Based on such network and attack models, we first study the link-level network robustness by investigating the link loss probability and the expected number of lost links. We then study the network-wide robustness under an area attack, where we first present four desired properties for a well-defined robustness measure and accordingly propose a new measure: the percentage of the surviving end-to-end communication pairs. Simulation results on a real-world network are given to verify our analytical results.

College Station, Texas, USA            Qing Zhou
California, USA                   Long Gao
China, People's Republic          Ruifang Liu
College Station Texas, USA        Shugang Cui

# Contents

# Chapter 1
# Introduction

To design a network that is robust under attacks, the first step is to define a robustness measure for evaluating the network response under attacks. Such a measure could be in different forms regarding different attacks. In the next, we first introduce several attack classifications: small-scale vs. large-scale attacks, physical vs. logical attacks, random vs. targeted attacks, and geographically concentrated vs. distributed attacks. Afterwards, we focus on a particular type of large-scale attacks, the *area attacks*, for which we investigate the appropriate robustness measure in this monograph.

## 1.1   Small-Scale Versus Large-Scale Attacks

Here, we first focus on two different attack classes: small-scale attacks vs. large-scale attacks.

### 1.1.1   Small-Scale Attacks

Small-scale attacks lead to single or a small amount of failures that are usually in the form of targeted losses of nodes or links, where the attackers attack on a small number of pre-determined nodes in order to manipulate certain aspects of the overall network functionality, such as the connectivity [1] or the maximum flow of the network [2, 3]. Specifically, for the connectivity-oriented one, the attackers target at maximizing the attack impact by tearing the network into disconnected components. For the maximum flow oriented ones, the objective for the attackers is to find an attack strategy to maximize certain utility functions, which leads to the Network Inhibition Problem (NIP) [4] or the Maximum Flow Network Interdiction Problem (MFNIP) [5]. In particular, the objective could be minimizing the maximum flow value in the network graph under the constraint that the total destruction cost is less than a fixed budget.

Q. Zhou et al., *Network Robustness under Large-Scale Attacks*, SpringerBriefs in Computer Science, DOI 10.1007/978-1-4614-4860-0_1, © The Authors 2013

## 1.1.2  Large-Scale Attacks

A massive amount failures occur under large-scale attacks, whose initial pattern could be either targeted or random. Besides the traditional targeted large-scale attacks, where a large amount of nodes or links are attacked simultaneously, a new effective form could be started by destroying a small amount of targeted nodes, such that a spontaneous chain reaction is triggered. Specifically, in a network system with highly dependent parts [6–8], the balance of loads could be destroyed by attacking a small fraction of connections, which triggers large-scale cascading failures across multiple interdependent parts [9–11]. On the other hand, large-scale attacks could happen in the form of random killings of a significant number of links or nodes in the network [12], where the function of the overall system leans on the percentage of surviving devices [13, 14]. It has been proved that the survived nodes are connected almost surely when the attack probability is below a certain critical value; otherwise the network is broken into disconnected parts with probability one [15, 16].

## 1.2  Physical Versus Logical Attacks

Another intuitive classification of attacks is to catalog them into physical attacks and logical attacks.

### 1.2.1  Physical Attacks

Under physical attacks, parts of the infrastructure are permanently destroyed and the damaged ones need to be physically replaced in order to recover the function of the network. One popular form of physical attacks is via bombing. The most effective bomb against communication networks is the Electromagnetic Pulse (EMP) bomb [17, 18]; by coupling with electrical systems, the resulting rapidly-changing electric and magnetic fields produce damaging current and voltage surges across the communications infrastructure. Another popular source for physical attacks is natural hazard, such as wild fires, earthquakes, hurricanes, or floods [19], where the resulting loss depends on the resilience of the corresponding networks. Besides the communication network, other vulnerable networks physical attacks are the transportation networks [20–22] and power grids [23, 24].

### 1.2.2  Logical Attacks

Logical attacks focus on manipulating networking software or information flows, for which the Internet has been the main target [15, 25], where wide-spread failures could be caused by simply inserting virus that spreads along the logical connections across the network [26–28]. Different from the permanent damages caused by physical attacks, logical attacks usually cause temporary malfunctions on the devices, where the system could be automatically recovered after the attacks are isolated and logically removed [29–31].

## 1.3 Random Versus Targeted Attacks

From the earlier discussion, we see that the attack patterns could also be divided into random attacks and targeted attacks, where the random attacks choose targets uniformly and the targeted attacks choose critical ones to destroy.

The uniform target selection for random attacks could be due to many reasons: e.g., the topology of the networks is unaccessible, where the attackers cannot figure out the distinctive values of different targets; the attack target is uncontrollable, where the attackers could only start an attack but cannot control the accuracy of the attack. Under a random attack, the robustness of a network is usually evaluated by some statistical measures, which is usually built upon some random network models. In particular, recent results on studying the World-Wide Web (WWW) [32–34], the Internet [35], and other large networks [36] indicate that many networks belong to an inhomogeneous network class known as scale-free networks [15]. One property of scale-free network models is that their connectivity distribution decays according to power-law. Meanwhile, scale-free networks show a high tolerance against random failures, where nodes fail with uniform probabilities. However, scale-free networks are extremely vulnerable under targeted attacks: When the hub nodes are attacked, the diameter of the network increases rapidly and many isolated fragments appear [37].

Targeted attacks try to damage certain components in a network, where people are interested in identifying the most vulnerable part of a given network to attack or to protect. For example, one recent study on the vulnerability of the fiber infrastructure proposed an algorithm to identify the most vulnerable part of the network by minimizing certain cost functions [38]. One way to make the system robust against such attacks is to add redundant components around the critical parts [39,40].

## 1.4 Geographically Concentrated Versus Distributed Attacks

Much of traditional research has been focusing on geographically distributed attacks, where the failures across physical neighbors are usually assumed independent [41], especially for the studies on logical attacks [42,43]. In recent years, more results appear on geographically concentrated failures over fiber-optic networks

[38, 44], wireless mesh networks [45, 46], and overlayed networks [47]. In such cases, the failures of the attacked nodes and links are correlated as a function of their geographic locations. Specifically, the attack on a particular node may affect its geographic neighbors with high probability, but may not affect its neighbors on the logic connectivity graph.

## 1.5   Robustness Measure for Area Attacks

In the rest of this monograph, we define and focus on a new type of attacks, i.e., the area attack, which is large-scale, random, physical, and geographically concentrated. As related works, in [38, 48], the authors studied large-scale geographically-concentrated failures in the USA fiber backbone network, where the focus is on identifying the most vulnerable parts of the fiber infrastructure under deterministic attacks, with the worst-case line segment and circular cuts being investigated as attack models. However, the results in these works are only applicable over a deterministic network topology with targeted attacks. We need to establish some fundamental results over random networks with random attacks, where identifying appropriate robustness measures is the focus in this monograph.

In earlier works, several measures have been proposed to evaluate the structural robustness of complex networks, e.g., the size of the giant component [12], the cluster size distribution [37], the network diameter [37], and the node connectivity [49]. However, as shown later, these measures may not be suitable for evaluating network performance under area attacks. For this new type of large-scale correlated attacks, many new questions rise naturally:

- How does the attack radius affect the link survivability?
- How many links will be destroyed in average by one attack?
- How much is the overall network functionality degraded by the attack?
- What kind of networks is more robustness to area attacks?

All the above questions point to the need of a good robustness measure for a network under area attacks. In the monograph, we first adopt the average probability of an arbitrary link being attacked and the expected number of destroyed links to study the link-level network robustness under area attacks. Then for the network-wide robustness, we first discuss four desired properties for a well-defined measure and accordingly propose a new measure based on the percentage of surviving end-to-end communication pairs. We provide the analytical and simulation results for both cases of the traditional random network and the random network with limited communication range (LCR), which show that a LCR random network is more robust compared with the traditional random network under area attacks.

The rest of the monograph is organized as follows. In Chap. 2, we describe the network and attack models for the area-attack problems. In Chap. 3, the link-level network robustness under area attacks is studied in terms of the average probability of an arbitrary link being attacked and the expected number of destroyed links for both the LCR random network model and the traditional random network model. In Chap. 4, we introduce a new measure of network robustness under area attacks and compare the robustness performance under the LCR random network model and the traditional random network model. The simulation results are given in Chap. 5 to verify our analytical results. Finally, we summarize our conclusions in Chap. 6.

# Chapter 2
# System Models

In this chapter, we describe the network models and the attack model for the area attack problem under consideration.

## 2.1 Network Models

We consider a geographic unit square area $\mathcal{W}$, i.e., $\mathcal{W} = \{(x,y) : 0 \leq x \leq 1, 0 \leq y \leq 1\}$. In the traditional random network model $G(\lambda, p_r)$ [50], the nodes are randomly distributed in $\mathcal{W}$ with density $\lambda$ and node pairs are connected randomly with a probability $p_r$.

In the LCR random network model $G_{LCR}(\lambda, p, l)$, the nodes are distributed randomly in the area $\mathcal{W}$ with density $\lambda$. The underlying connectivity graph is formed as: If two nodes are located within a distance $l$, they are connected to each other with a probability $p$; there are no connections between nodes that are located further than $l$-distance apart. We also assume the periodic boundary condition (PBC) [51] on $\mathcal{W}$ in the LCR random network model to eliminate the boundary effects, i.e., if a node is located at $(x,y)$, we assume that the node also appears at $(x+n, y+m)$ with $n, m \in \mathbb{Z}$. The effect of PBC simplification will be later discussed. The difference between the LCR random network model and the traditional random network model is that the former one has a range limitation and the latter does not, such that the PBC is not suitable for the traditional random network model. We assume that $p$ and $p_r$ are large enough to maintain a full connectivity for both network models.

## 2.2 Attack Model

We model the attack area as a circular disk with radius $r$ and center $\mathbf{c}(x_\mathbf{c}, y_\mathbf{c})$, where $\mathbf{c}$ is uniformly distributed within $\mathcal{W}$. We refer to the region being attacked as $\mathscr{A}(\mathbf{c}, r)$, in which all the nodes and links are destroyed. We demonstrate an example of the

Q. Zhou et al., *Network Robustness under Large-Scale Attacks*, SpringerBriefs in
Computer Science, DOI 10.1007/978-1-4614-4860-0_2, © The Authors 2013

**Fig. 2.1** A network example,
$\lambda = 100, l = 0.2, p = 0.5$, and
$r = 0.1$

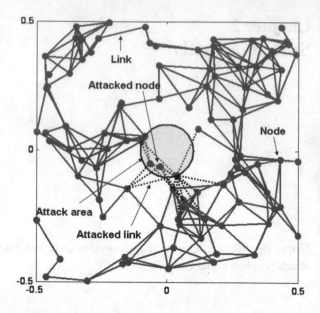

LCR random network under an area attack in Fig. 2.1 ($\lambda = 100$, $l = 0.2$, $p = 0.5$, and $r = 0.1$), where the disk in the center is the attack area, with the dotted line segments and shadowed nodes touched by the attack area being the destroyed links and nodes.

# Chapter 3
# Link-Level Network Robustness to Area Attacks

In this chapter, we study the link-level network robustness to area attacks. We first derive the average probabilities of an arbitrary link being attacked for both the LCR random network and the traditional random network. Afterwards, we present the expected numbers of the destroyed links for both cases.

## 3.1 Average Probability of an Arbitrary Link being Attacked

In this section, we first present the probability of being attacked for a link with a given length. Then we derive the average loss probability for an arbitrary link for both the LCR random network case and the traditional random network case.

As shown in Fig. 3.1 with a given attack area $\mathscr{A}(\mathbf{c}, r)$, for a particular link $L_1$ with length $l_1$, it is destroyed if the center of the attack area is in the olivary area $S$. Therefore, the probability of being attacked for $L_1$ can be expressed as

$$P_{\text{attack}}(l_1) = \frac{|S|}{|\mathscr{W}|} = \frac{(2l_1 r + \pi r^2)}{1} = 2l_1 r + \pi r^2, \tag{3.1}$$

where $|\cdot|$ denotes the volume.

Since the nodes and the attack area are uniformly distributed, the network geometry has a symmetric property, such that the probability of a link being killed is the same as other links of the same length (by neglecting the boundary effects).

### 3.1.1 The LCR Random Network Case

In a LCR random network, we assume the PBC on $\mathscr{W}$. We denote $f_{LCR}(x)$ as the probability density function (PDF) of the link length, and we have

$$f_{LCR}(x) \cdot dx \tag{3.2}$$

Q. Zhou et al., *Network Robustness under Large-Scale Attacks*, SpringerBriefs in Computer Science, DOI 10.1007/978-1-4614-4860-0_3, © The Authors 2013

**Fig. 3.1** Single link example

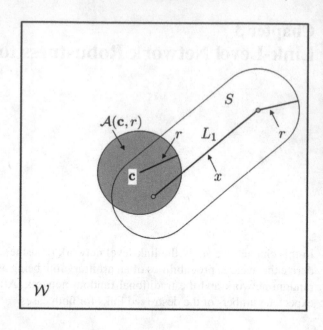

$$= Pr(\text{A randomly selected link has the length between } x \text{ and } x+dx) \quad (3.3)$$

$$\triangleq P_{LCR}(x) \quad (3.4)$$

Since we assume the PBC along both the $x$ and $y$ directions, the network is homogeneous. Let us fix one end of a link; and $P_{LCR}(x)$ is equal to the probability of the other end being located in the ring area centered at the first end along the radius from $x$ to $x+dx$ [52], which could be expressed as

$$P_{LCR}(x) = \frac{2\pi x \cdot dx}{\pi l^2}, \quad (3.5)$$

where $2\pi x \cdot dx$ is the area of the ring and $\pi l^2$ is the area of the disk with the radius of $l$.

Therefore, we have

$$f_{LCR}(x) = \frac{2}{l^2}x, \qquad x \in [0,l]. \quad (3.6)$$

Combining (3.1) and (3.6), the average probability of an arbitrary link being attacked in the LCR random network model is given by

$$P_{LCR} = \int P_{attack}(x) f(x) dx \quad (3.7)$$

$$= \frac{4}{3}lr + \pi r^2. \quad (3.8)$$

### 3.1.2 The Traditional Random Network Case

In the traditional random network, there is no local limitation on the link length and the range of a link length is $(0, \sqrt{2})$. We denote $f_{RN}(x)$ as the PDF of the link length in the traditional random network. Similar to the LCR random network case, the average probability of an arbitrary link being attacked in the traditional random network model is given by

$$P_{RN} = \int P_{attack}(x) f_{RN}(x) dx \qquad (3.9)$$

$$= \frac{4 + 2\sqrt{2}}{15} r + \pi r^2 - \frac{2}{3} r \log(1 + \sqrt{2} - \frac{4}{3} r) \log(\sqrt{2} - 1), \qquad (3.10)$$

where $f_{RN}(x)$ is derived in Appendix and is given by

$$f_{RN}(x) = \begin{cases} 2\pi x - 8x^2 + 2x^3, & x \in [0,1] \\ 2x(-2\arcsin\left(\frac{-2+x^2}{x^2}\right) - 2 + 4(x^2 - 1)^{\frac{1}{2}} - x^2), & x \in \left(1, \sqrt{2}\right) \end{cases} \qquad (3.11)$$

## 3.2 The Expected Number of Destroyed Links

In this section, we first derive the expected number of destroyed links in an arbitrary LCR random network. We then present the result in an arbitrary traditional random network. Finally, we investigate the case when the network topology is fixed.

### 3.2.1 The LCR Random Network Case

In the LCR random network, the expected total number of node pairs is given by

$$N_{pair} = E(C_2^N) = \frac{\lambda^2}{2} \qquad (3.12)$$

where $N$ is a Poisson random variable with density $\lambda$.

The probability that a link exists between two randomly chosen nodes $n_i$ and $n_j$ can be expressed as

$$P(||n_i - n_j|| \leq l) = \frac{\pi l^2}{|\mathscr{W}|} p = \pi l^2 p, \qquad (3.13)$$

where $||n_i - n_j||$ is the distance between $n_i$ and $n_j$.

**Table 3.1** Notations in the proof of Theorem 3.1

| Terms | Definition |
|-------|------------|
| $\Omega_A$ | The sample space of all possible realizations of the area attack |
| $A$ | A random variable which is drawn from $\Omega_A$ with the same probability |
| $A_m$ | An arbitrary area attack in $\Omega_A$ |
| $\Omega_N$ | The sample space of all possible network realizations |
| $N$ | A random network which is drawn from $\Omega_N$ with the same probability |
| $N_i$ | An arbitrary network realization in $\Omega_N$ |
| $\Omega_L$ | The sample space of all possible realizations of the links |
| $L$ | A random link which is drawn from $\Omega_L$ with the same probability |
| $L_N$ | The set of the links in the network $N$ |
| $l_{j,N}$ | An arbitrary link in $L_N$ |
| $\mathscr{P}_N$ | The set of all paths which connect all the communication pairs in the given network $N$ |
| $p_{j,N}$ | An arbitrary path in $\mathscr{P}_N$ |

Therefore, the expected number of links in the network can be calculated as

$$N_{LCR} = N_{pair}P(||n_i - n_j|| \leq l) = \frac{1}{2}\pi l^2 p\lambda^2. \qquad (3.14)$$

**Theorem 3.1.** *Although the attacks over different links may be correlated, the expected number of destroyed links in the whole network is given as*

$$N_{d\_LCR} = N_{LCR}P_{LCR}, \qquad (3.15)$$

*which is the same as the case when each link gets attacked independently.*

*Proof.* First, we define some notations in Table 3.1.

We assume that $L$ and $A$ are independent, and $N$ and $A$ are independent. Let $1_{\{l_{j,N},A\}}$ denotes an indicator function given as:

$$1_{\{l_{j,N},A\}} = \begin{cases} 1, & \text{the link } l_{j,N} \text{ is attacked by } A \\ 0, & \text{otherwise} \end{cases} \qquad (3.16)$$

The number of the destroyed links in $N$ by an area attack $A$ can be expressed as

$$\sum_{l_{j,N}\in L_N} 1_{\{l_{j,N},A\}}. \qquad (3.17)$$

Therefore the expected number of destroyed links in the network under the area attack is given as

$$E_{N,A}\left(\sum_{l_{j,N}\in L_N} 1_{\{l_{j,N},A\}}\right), \qquad (3.18)$$

which implies that, we count the number of the destroyed links for every possible network realization and every possible area attack, and then average the number over the all the possible network realizations and all the possible area attacks.

On the other hand, the expected number of links in the network can be expressed as $E_N(|L_N|)$. Therefore, we have

$$E_{N,A}\left(\sum_{l_{j,N}\in L_N} 1_{\{l_{j,N},A\}}\right) \tag{3.19}$$

$$= E_N\left(E_A\left(\sum_{l_{j,N}\in L_N} 1_{\{l_{j,N},A\}}\right)\right) \tag{3.20}$$

$$= E_N\left(\sum_{l_{j,N}\in L_N} E_A\left(1_{\{l_{j,N},A\}}\right)\right) \tag{3.21}$$

$$= E_N\left(\sum_{l_{j,N}\in L_N} P_{attack}\left(x_{l_{j,N}}\right)\right) \tag{3.22}$$

$$= \frac{\sum_{N\in\Omega_N}\sum_{l_{j,N}\in L_N} P_{attack}\left(x_{l_{j,N}}\right)}{|\Omega_N|} \tag{3.23}$$

$$= \frac{\lim_{\max\Delta_i\to 0}\sum_{i=0}^{n}\left(\left(\sum_{N\in\Omega_N}|L_N|\right)\cdot f_{RN}(t_i)\,\Delta_i\cdot P_{attack}(t_i)\right)}{|\Omega_N|}, \tag{3.24}$$

where $x_{l_{j,N}}$ is the length of $l_{j,N}$, $0 = a_0 \le t_1 \le a_1 \le t_2 \le a_2 \le \cdots \le a_{n-1} \le t_n \le a_n = l$ is a tagged partition over the range of the link length, $\Delta_i = a_i - a_{i-1}$ is the width of sub-interval $i$, $\sum_{N\in\Omega_N}|L_N|$ is the total number of links in all network realizations, and $f_{RN}(t_i)\Delta_i$ is the probability of the length of an arbitrary link being in $(a_{i-1}, a_i)$. The Riemann sums in (3.24) can be written as

$$\frac{\int \sum_{N\in\Omega_N}|L_N|\cdot f_{RN}(x)\cdot P_{attack}(x)dx}{|\Omega_N|} \tag{3.25}$$

$$= \frac{\sum_{N\in\Omega_N}|L_N|}{|\Omega_N|}\cdot\int f_{RN}(x)\cdot P_{attack}(x)dx \tag{3.26}$$

$$= E_N(|L_N|)\cdot\int f_{RN}(x)\cdot P_{attack}(x)dx, \tag{3.27}$$

As such, we prove that $N_{d\_LCR} = N_{LCR}P_{LCR}$, i.e., the expected number of destroyed links under area attack is the same as the one under link-independent attacks.

From the above theorem it is clear that, in general, the attacks over different links are correlated in each area attack realization. However, the expected number of attacked links, which is averaged over all realizations, is only dependent on $P_{LCR}$; the attack correlation between the links is eliminated by the expectation operation.

### 3.2.2  The Traditional Random Network Case

Given the fact that the expected total number of links in the traditional random network is

$$N_{RN} = E(C_2^N)p_r = \frac{1}{2}\lambda^2 p_r, \tag{3.28}$$

the expected number of destroyed links by a randomly located area attack can be calculated as

$$N_{d\_RN} = N_{RN}P_{RN}. \tag{3.29}$$

The above simple relation follows a similar argument to that of Theorem 3.1.

### 3.2.3  The Fixed Network Case

In the previous analysis of the section, all the measures are expected values over all possible random realizations of the network topology. In the next, we fix a network realization, i.e., the geographic locations of all the links and the corresponding connectivity graph are fixed. We then derive the expected number of destroyed links only over the random realizations of the attack area. The following result is applicable to analyzing the performance of a given deterministic network under area attacks.

Let $M$ be the number of destroyed links in a given network. The expectation of $M$ over attack locations can be expressed as

$$E(M) = \sum_{i=1}^{\infty} iq_i, \tag{3.30}$$

where $q_i$ denotes the probability of a total of $i$ links being attacked.

For a given link, we know that the link is attacked if the center of the attack area is within the corresponding olivary region around the link as shown in Fig. 3.1. Therefore, $q_i$ can be expressed as

$$q_i = \frac{|S_i|}{|\mathscr{W}|}, \tag{3.31}$$

where $|S_i|$ represents the sum area of all the overlapping parts intersected exactly by $i$ olivary regions around all sets of $i$ links. With the example shown in Fig. 3.2, for $i = 1$, $|S_1|$ denotes the sum area of the shadowed parts that each covers exactly one link, i.e., $|S_1| = |S_{11}| + |S_{12}| + |S_{13}| + |S_{14}|$; for $i = 2$, $|S_2| = |S_{21}| + |S_{22}| + |S_{23}| + |S_{24}|$; and for $i = 3$, $|S_3| = |S_{31}|$.

**Fig. 3.2** An example of the overlapping olivary regions

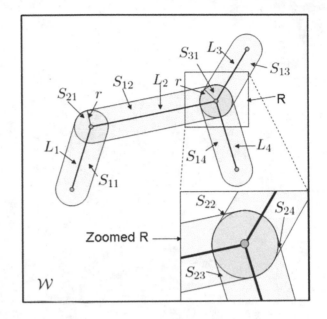

Substituting (3.31) into (3.30), we have

$$E(M) = \frac{\sum_{n=1}^{N} |T_n|}{|\mathcal{W}|}, \tag{3.32}$$

where $N$ is the total number of links and $T_n$ denotes the area of the olivary region around link $n$.

For the example in Fig. 3.2, we now illustrate the details on how to calculate (3.32). Specifically, we have

$$|T_1| = |S_{11}| + |S_{21}|, \tag{3.33}$$

$$|T_2| = |S_{12}| + |S_{21}| + |S_{22}| + |S_{23}| + |S_{31}|, \tag{3.34}$$

$$|T_3| = |S_{13}| + |S_{22}| + |S_{24}| + |S_{31}|, \tag{3.35}$$

$$|T_4| = |S_{14}| + |S_{23}| + |S_{24}| + |S_{31}|. \tag{3.36}$$

Therefore, the expectation (over attack locations) of $M$ in this given network realization can be expressed as

$$E(M) = \sum_{i=1}^{3} i q_i$$

$$= 1 \cdot \frac{|S_1|}{|\mathcal{W}|} + 2 \cdot \frac{|S_2|}{|\mathcal{W}|} + 3 \cdot \frac{|S_3|}{|\mathcal{W}|}$$

$$= \frac{|T_1| + |T_2| + |T_3| + |T_4|}{|\mathcal{W}|}. \tag{3.37}$$

# Chapter 4
# Network-Wide Robustness to Area Attacks

In this chapter, we consider the network-wide robustness to area attacks. We first give four properties that a good network robustness measure should possess. We then propose a new measure for network robustness: the percentage of surviving communication pairs, which can be proven to satisfy the desired four properties. We further compare other existing network robustness measures against the proposed one. Afterwards, we analyze the robustness of the LCR random network and the traditional random network.

## 4.1 Properties of a Good Robustness Measure

Before discussing the properties that a good network robustness measure should have, we first define the segment size vector. Let $\mathbf{S} = (s_1, s_2, \dots, s_k)$ denotes a segment size vector for a given network, where $s_i$ is the number of nodes in the $i$th segment and $k$ is the total number of segments. For a network $G$ under attack, the robustness measure $R$ should at least have the following four properties.

*Property 4.1 (Normalization).* For the convenience of comparison, we desire $0 \leq R \leq 1$, where a higher $R$ value implies a more robust network.

This property makes the robustness of two different networks comparable.

*Property 4.2 (Permutation).* If $G_1$ and $G_2$ are two networks with $\mathbf{S}_1 = (s_1, s_2, \dots, s_k)$ and $\mathbf{S}_2 = (s_{m_1}, s_{m_2}, \dots, s_{m_k})$, such that $m_i \in \{1, 2, \dots, k\}$, $i \in \{1, 2, \dots, k\}$, and $m_i \neq m_j, \forall i \neq j$, we have $R(\mathbf{S}_1) = R(\mathbf{S}_2)$.

The permutation property indicates that the measure value $R$ does not change with the order of the segments in $\mathbf{S}$, which is an intuitive property that a good network robustness measure should have.

*Property 4.3 (Biasing).* If $G_1$ and $G_2$ are two networks with $\mathbf{S}_1 = (s_1, s_2, \dots, s_i, \dots, s_j, \dots, s_k)$ and $\mathbf{S}_2 = (s_1, s_2, \dots, s_i', \dots, s_j', \dots, s_k)$, such that $s_i + s_j = s_i' + s_j'$ and $|s_i - s_j| > |s_i' - s_j'|$, we have $R(\mathbf{S}_1) > R(\mathbf{S}_2)$.

Q. Zhou et al., *Network Robustness under Large-Scale Attacks*, SpringerBriefs in Computer Science, DOI 10.1007/978-1-4614-4860-0_4, © The Authors 2013

With the biasing property, a network with a dominant segment has a greater value of $R$ than the one with the same number of nodes but several equal-size segments, which is based on the expectation that a robust network should lead to a dominate segment instead of several similar-size prices after an attack.

*Property 4.4 (Splitting).* If $G_1$ and $G_2$ are two networks with $\mathbf{S_1} = (s_1, s_2, \ldots, s_i, \ldots, s_k)$ and $\mathbf{S_2} = (s_1, s_2, \ldots, s_{i1}, s_{i2}, \ldots, s_k)$, such that $s_i = s_{i1} + s_{i2}$, we have $R(\mathbf{S_1}) > R(\mathbf{S_2})$.

The splitting property implies that $R$ decrease if one segment splits into two segments in a network, which is based on the expectation that a robust network should lead to a less number of segments after an attack.

## 4.2   The Percentage of Surviving Communication Pairs

For a network under attacks, if there is at least one multihop path that connects two nodes, which means there two nodes can still communicate to support a certain application, we define them as a communication pair. In other words, if these two nodes are in the same network segment, we say that they construct a communication pair. Therefore, if we assume that the network is fully connected before the attack, the percentage of surviving communication pairs after the attack can be expressed as

$$C(\mathbf{S}) = \frac{\sum_{i=1}^{k} C_2^{s_i}}{C_2^{N}},  \tag{4.1}$$

where $N$ is the total number of nodes in the given network; $C_y^x$ is the number of combinations from "$x$ choose $y$" such that $C_2^{s_i}$ is the total number of communication pairs in the $i$th segment and $C_2^N$ is the total number of communication pairs in the network before the attack.

In the following, we prove that the above measure in (4.1) satisfies all the four desired properties introduced in Sect. 4.1. First of all, the proofs of satisfying Property 4.1 and Property 4.2 are straightforward based on the definition in (4.1).

*Property 4.3* (Biasing):

*Proof.* Given $\mathbf{S_1}$ and $\mathbf{S_2}$ such that $s_i + s_j = s_i' + s_j'$ and $|s_i - s_j| > |s_i' - s_j'|$, we have

$$C(\mathbf{S_1}) - C(\mathbf{S_2})$$

$$= \frac{C_2^{s_i} + C_2^{s_j}}{C_2^{N}} - \frac{C_2^{s_i'} + C_2^{s_j'}}{C_2^{N}}$$

$$= \frac{(s_i - s_j)^2 - (s_i' - s_j')^2 + (s_i + s_j)^2 - (s_i' + s_j')^2}{2C_2^{N}}$$

$$> 0,  \tag{4.2}$$

which completes the proof.

*Property 4.4* (Splitting):

*Proof.* Given $\mathbf{S_1}$ and $\mathbf{S_2}$ such that $s_i = s_{i1} + s_{i2}$, we have

$$C(\mathbf{S_1}) - C(\mathbf{S_2})$$

$$= \frac{C_2^{s_i}}{C_2^N} - \frac{C_2^{s_{i1}} + C_2^{s_{i2}}}{C_2^N}$$

$$= \frac{s_i^2 - s_i - (s_{i1}^2 - s_{i1} + s_{i2}^2 - s_{i2})}{2C_2^N} > 0, \qquad (4.3)$$

which completes the proof.

## 4.3   Comparison Against Existing Measures

There are several existing measures on network robustness in the literature: the giant segment size $\tilde{S}$ [37], the average segment size $\bar{s}$ [37], the diameter of the network $d$ [37], and the average shortest path length $\bar{l}_{path}$ [53]. We will show that there existing measures cannot satisfy all the four properties that we discussed earlier, which are necessary for being a good network robustness measure.

Specifically, the giant segment size $\tilde{S}(\mathbf{S})$ is defined as the number of nodes in the largest segment, i.e., $\tilde{S}(\mathbf{S}) = max\{s_1, s_2, \ldots, s_k\}$. It is easy to verify that the biasing and splitting properties cannot hold for $\tilde{S}(\mathbf{S})$. Intuitively, from $\tilde{S}(\mathbf{S})$, we only know that all the other segment sizes are less than $\tilde{S}(\mathbf{S})$ without the knowledge of local connectivity robustness in the other segments, which may lead to a wrong inference on the overall network robustness. In particular, under an attack a network might be broken into many segments of similar sizes: for example, the exponential network under random node attacks [37]. In such a scenario, the percentage of surviving communication pairs could represent the overall network robustness at any different network scale, while the giant segment size could not. Quantitatively, $\tilde{S}(\mathbf{S})$ and $C(\mathbf{S})$ are related as

$$C(\mathbf{S}) \geq \frac{C_2^{\tilde{S}(\mathbf{S})}}{C_2^N}. \qquad (4.4)$$

In general, the above bound is quite loose, which implies that $\tilde{S}(\mathbf{S})$ and $C(\mathbf{S})$ carry dramatically different amounts of information regarding node-to-node connectivities in the whole network. Especially when the sizes of other segments are on the same order of $\tilde{S}(\mathbf{S})$, $C(\mathbf{S})$ could be several times larger than $\frac{C_2^{\tilde{S}(\mathbf{S})}}{C_2^N}$. On the other hand, when the above bound is tight, $\tilde{S}(\mathbf{S})$ and $C(\mathbf{S})$ carry similar messages since from one we could learn about the other.

Next, we consider the average segment size $\bar{s}(\mathbf{S})$, which is defined as the average size of the $k$ segments in the network, i.e.,

$$\bar{s}(\mathbf{S}) = \frac{1}{k}\sum_{i=1}^{k} s_i = \frac{N}{k}. \tag{4.5}$$

It is easy to verify that the biasing property cannot hold for $\bar{s}(\mathbf{S})$. Similar to $\tilde{S}(\mathbf{S})$, from $\bar{s}(\mathbf{S})$ we cannot get the knowledge of local connectivity robustness in all the segments. Quantitatively, the relation between $\bar{s}(\mathbf{S})$ and $C(\mathbf{S})$ is given by

$$C(\mathbf{S}) \geq \frac{kC_2^{\bar{s}(\mathbf{S})}}{C_2^N}, \tag{4.6}$$

where the equality holds when all segments have the same size. This relation could be proved by using Property 4.2 and Property 4.3, sketched as follows.

*Proof.* Let $\mathbf{S}^0 = (s_1^0, s_2^0, \ldots, s_k^0)$ denote the segment size vector with an increasing order of the segment sizes, i.e., $s_1^0 \leq s_2^0 \leq \ldots \leq s_k^0$, with $(s_1^0, s_2^0, \ldots, s_k^0)$ being a permutation of $(s_1, s_2, \ldots, s_k)$. By Property 4.2, we have

$$C(\mathbf{S}) = C(\mathbf{S}^0). \tag{4.7}$$

At time $i$, we select the two segments with the smallest size and the biggest size in $\mathbf{S}^i$, and average them as two equal-size segments. We sort and name the newly generated segment vector with an increasing order of the segment size as $\mathbf{S}^{i+1}$. By Property 4.3, we have

$$C(\mathbf{S}^i) > C(\mathbf{S}^{i+1}). \tag{4.8}$$

We continue this resizing process until time $k$, when all the segments have the same size $\bar{s}(\mathbf{S})$. Therefore, we have

$$C(\mathbf{S}) = C(\mathbf{S}^0) > C(\mathbf{S}^1) > \ldots > C(\mathbf{S}^k) = \frac{kC_2^{\bar{s}(\mathbf{S})}}{C_2^N}. \tag{4.9}$$

Another existing network robustness measure is the network diameter $d$, which is defined by the longest shortest path between any pair of nodes of the network, i.e.,

$$d(\mathbf{S}) = \max_{0 \leq i,j \leq k}(d_{ij}), \tag{4.10}$$

where $d_{ij}$ is the shortest path between node $i$ and node $j$. It is easy to verify that the biasing and splitting properties cannot hold for $d$. Intuitively, all the paths between a particular pair of nodes may not survive after the attack, which leads to an infinitely large $d$. Thus, the diameter measure is not suitable to measure a fragmented network.

**Table 4.1** A comparison between the proposed measure and other existing measures

|  | Normalization | Permutation | Biasing | Splitting |
|---|---|---|---|---|
| $R$ | ✓ | ✓ | ✓ | ✓ |
| $\tilde{S}$ | ✗ | ✓ | ✗ | ✗ |
| $\bar{s}$ | ✗ | ✓ | ✗ | ✓ |
| $d$ | ✗ | ✓ | ✗ | ✗ |
| $\bar{l}_{path}$ | ✗ | ✓ | ✗ | ✗ |

Finally, we consider the average shortest path $\bar{l}_{path}$, which is the average of the shortest distances between every pair of nodes, i.e.,

$$\bar{l}_{path}(\mathbf{S}) = \frac{\sum_{0 \leq i,j \leq k} d_{ij}}{C_2^N}. \tag{4.11}$$

It is easy to verify that the biasing and splitting properties cannot hold for $\bar{l}_{path}(\mathbf{S})$. Similar to $d$, $\bar{l}_{path}(\mathbf{S})$ is not suitable to measure a fragmented network.

A comparison between the proposed measure and other existing measures is given in Table 4.1. The proposed measure is the only one that meets all the four properties, which makes it a good candidate to measure the network robustness under area attacks. In the next section, we will compare the robustness of the LCR random network against the traditional random network in terms of the proposed measure.

## 4.4 Robustness Study

In this section, we study the robustness of both the LCR and traditional random networks with the measure proposed in Sect. 4.2. We first prove that the percentage of the surviving communication pairs is equal to the average probability that an arbitrary communication pair survives. We then adopt the latter to conduct the robustness study. Since such a probability is hard to quantify in general, here we only derive the lower and upper bounds. The exact analysis is left for our future work.

**Theorem 4.1.** *The percentage of the surviving communication pairs is equal to the average probability that an arbitrary communication pair survives, i.e.,*

$$C = \overline{\Pr}(An\ arbitrary\ communication\ pair\ survives) \triangleq P_{CPS}. \tag{4.12}$$

*Proof.* Similar to the proof of Theorem 3.1, the notations used in this proof can be found in Table 3.1.

We assume that $\mathscr{P}_N$ and $A$ are independent, and $N$ and $A$ are independent. Let $1_{\{p_{j,N},A\}}$ denotes an indicator function as

$$1_{\{p_{j,N},A\}} = \begin{cases} 1, & \text{the path } p_{j,N} \text{ is not attacked by } A \\ 0, & \text{otherwise} \end{cases} \tag{4.13}$$

The percentage of the surviving communication pairs in a given $N$ can be calculated as

$$\frac{\sum_{p_{j,N}\in\mathscr{P}_N} 1_{\{p_{j,N},A\}}}{|\mathscr{P}_N|}. \tag{4.14}$$

Therefore, the percentage of the surviving communication pairs averaged over all possible network and area attack realizations and can be expressed as

$$C = \frac{\sum_{N\in\Omega_N}\sum_{A\in\Omega_A}\frac{\sum_{p_{j,N}\in\mathscr{P}_N} 1_{\{p_{j,N},A\}}}{|\mathscr{P}_N|}}{|\Omega_N|\cdot|\Omega_A|}. \tag{4.15}$$

The probability that a communication pair that is connected by path $p_{j,N}$ survives in $N$ can be calculated as

$$P_{p_{j,N}} \triangleq \frac{\sum_{A\in\Omega_A} 1_{\{p_{j,N},A\}}}{|\Omega_A|}. \tag{4.16}$$

The average probability that an arbitrary communication pair survives can be expressed as

$$P_{CPS} = E_{N,\mathscr{P}_N}(P_{p_{j,N}}) \tag{4.17}$$

$$= E_N(E_{\mathscr{P}_N}(P_{p_{j,N}})) \tag{4.18}$$

$$= \frac{\sum_{N\in\Omega_N}\frac{\sum_{p_{j,N}\in\mathscr{P}_N}\frac{\sum_{A\in\Omega_A} 1_{\{p_{j,N},A\}}}{|\Omega_A|}}{|\mathscr{P}_N|}}{|\Omega_N|} \tag{4.19}$$

$$= \frac{\sum_{N\in\Omega_N}\sum_{A\in\Omega_A}\frac{\sum_{p_{j,N}\in\mathscr{P}_N} 1_{\{p_{j,N},A\}}}{|\mathscr{P}_N|}}{|\Omega_N|\cdot|\Omega_A|} \tag{4.20}$$

$$= C,$$

where the derivation from (4.19) to (4.20) is due to the fact that $A$ is independent of $\mathscr{P}_N$. Therefore, we proved that $C$ is equal to $P_{CPS}$.

In the following, we first derive a universal (applicable to both of the two random network models) upper bound of the average probability that an arbitrary communication pair survives after the attack, which can be obtained by assuming that all the nodes outside the attack area are fully connected. The probability of the node being out of the attack area is $1 - \pi r^2$. Thus, the upper bound of the probability that a communication pair survives after the attack is the probability that the two ending nodes in the communication pair are both outside the attack area, i.e.,

$$P_{CPS} \leq p_{upper} = (1 - \pi r^2)^2. \tag{4.21}$$

The lower bound of the probability that an arbitrary communication pair survives needs to be investigated separately for each of the two network models. In the LCR random network model, the attack area $\mathscr{A}(\mathbf{c}, r)$ only affects the disk area $\mathscr{D}(\mathbf{c}, r+l)$ with the center $\mathbf{c}$ and radius $r+l$. Given the assumption that the network before the attack is fully connected with the argument of percolation (when $\lambda$ and $p$ are large enough), the left-over network outside region $\mathscr{D}(\mathbf{c}, r+l)$ is still fully-connected with high probability [54]. Since the probability that one node being out of $\mathscr{D}(\mathbf{c}, r+l)$ is $1 - \pi(r+l)^2$, the lower bound of the probability that a communication pair survives after the attack is the probability that the two ending nodes are both outside $\mathscr{D}$, i.e.,

$$P_{CPS}^{LCR} \geq p_{lower}^{LCR} = (1 - \pi(r+l)^2)^2. \tag{4.22}$$

In order to obtain the lower bound for the random network case, we take an arbitrary path that connects two nodes, and assume that the communication pair is destroyed as long as the path is destroyed by the attack. Like in Sect. 3.1, the probability that the path is destroyed can be expressed as

$$p_{path}(l_{path}) = \pi r^2 + 2l_{path}r, \tag{4.23}$$

where $l_{path}$ is the length of the path.

Therefore, the lower bound of the probability that a link with length $l_{path}$ survives after the attack for the random network case is given by

$$p_{lower}^{RN}(l_{path}) = 1 - p_{path}(l_{path}) = 1 - \pi r^2 - 2l_{path}r. \tag{4.24}$$

Thus, the lower bound of the average probability that an arbitrary link survives after the attack can be written as

$$P_{CPS}^{RN} \geq p_{lower}^{RN} = E(p_{lower}(l_{path})) = 1 - \pi r^2 - 2E(l_{path})r, \tag{4.25}$$

with

$$E(l_{path}) = 0.5214 \times \frac{\log \lambda}{\log \lambda p}, \tag{4.26}$$

where 0.5214 is the expected link length in the traditional random network model (see the proof in the Appendix), and $\frac{\log \lambda}{\log \lambda p}$ is the average number of hops for a particular communication pair connection when $\lambda$ is large [55].

# Chapter 5
# Simulation Results

In this chapter, we present the simulation results of this monograph. In the first part, we present the simulation results to evaluate both the link-level and network-wide robustness of the LCR random network under area attacks and compare it against that of the traditional random network. In the second part, we show the simulation results on a fiber plant of a major network provider [56]. We first assume that the topology information of this fiber network is known and show the expected number of attacked links by using the theoretical result of the fixed network case. Then without knowing the detail information about this fiber network but only some high-level statistical information, such like the node number, the link number, and the link length, we present the results by assuming that the fiber network is a LCR random network and a traditional random network, respectively.

In this chapter, the traditional random network $G(\lambda, p_r)$ is assumed to have the same density $\lambda$ with the LCR random network, and the connection probability $p_r$ is set to be

$$p_r = p\pi l^2, \tag{5.1}$$

which makes the expected numbers of links for both network models the same in order to have a fair comparison.

In Fig. 5.1, we draw the average probability of an arbitrary link being attacked for different values of attack area radius in both LCR and traditional random network cases. The system parameters are set as $\lambda = 180$, $l = 0.15$, and $p = 0.2, 0.8$. In the figure, the theoretical results for LCR random networks and traditional random networks are based on the results given in (3.8) and (3.10). The simulation results are obtained as follows: We first generate a large number of network realizations; for each network realization, we simulate a large number of area attacks; and the results are obtained by averaging the numbers of destroyed links over all realizations of network and area attacks. There are several observations from this figure: (i) The average probability of an arbitrary link being attacked in the LCR random network is roughly one half of that in the traditional random network, which means that the LCR network is more robustness than the traditional random network under area attacks. This is because that the traditional random network has more long links

Q. Zhou et al., *Network Robustness under Large-Scale Attacks*, SpringerBriefs in Computer Science, DOI 10.1007/978-1-4614-4860-0_5, © The Authors 2013

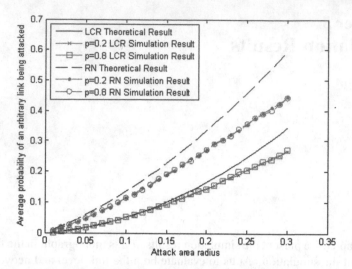

**Fig. 5.1** Average loss probability for an arbitrary link, $p = 0.2, 0.8$

than the LCR random network, which are more vulnerable to area attacks; (ii) the average probabilities of an arbitrary link being attacked in both cases are on the order of $r^2$ and are independent of the link connection probability $p$, which matches the theoretical results given in (3.8) and (3.10); (iii) there exists a gap between the simulation result and the theoretical result, which is caused by the boundary effect: When the center of the attack area is close to the boundary of W, part of the attack area is outside W, which is still counted in (3.1) but not considered in the calculation of $P_{LCR}$ and $P_{RN}$.

In Fig. 5.2, we draw the theoretical and simulation results for the expected number of attacked links in the LCR random networks based on (3.15), where the simulation parameters are set as $\lambda = 180$, $l = 0.15$, and $p = 0.2, 0.5, 0.8$. From the figure, we see that there is a gap between the theoretical and simulation results, which is caused by the boundary effect as we discussed above. In Fig. 5.3, we draw the theoretical and simulation results for the expected number of attacked links in the traditional random networks based on (3.29). The simulation parameters are the same as in Fig. 5.2. Compared with the results in Fig. 5.2, the expected numbers of attacked links in the traditional random networks are much larger than the ones in the LCR random networks.

In Fig. 5.4, we draw the theoretical and simulation results for the expected number of attacked links in one realization of the LCR random network. The simulation parameters are the same as in Fig. 5.2. In this figure, the gap between the theoretical and simulation results is also caused by the boundary effect in the theoretical calculation: When a link is close to the boundary of $\mathscr{W}$, part of the olivary region around the link is outside $\mathscr{W}$, which is still counted in the theoretical calculation of $T_n$ in (3.32).

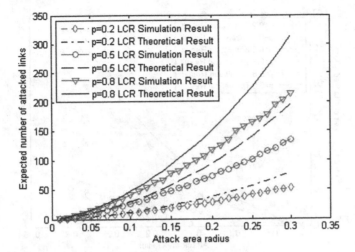

**Fig. 5.2** Expected number of attacked links in LCR random networks, $\lambda = 180, l = 0.15, p = 0.2, 0.5, 0.8$

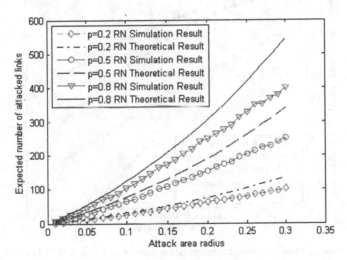

**Fig. 5.3** Expected number of attacked links in the traditional random network, $\lambda = 180, l = 0.15, p = 0.2, 0.5, 0.8$

In Fig. 5.5, we draw the simulated percentage of surviving communication pairs after an area attack, along with the theoretical upper and lower bounds, where the simulation parameters are set as $\lambda = 180$, $p = 0.35$, and $l = 0.15$. From the figure, we see that the simulation result for the LCR random network is close to the upper

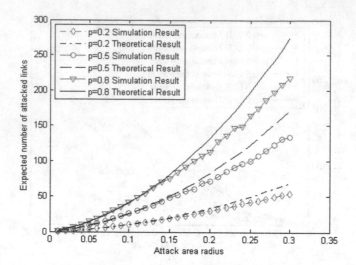

**Fig. 5.4** Expected number of attacked links in a given network

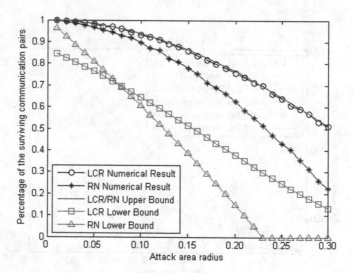

**Fig. 5.5** Percentage of the surviving communication pairs, $\lambda = 180, p = 0.35, l = 0.15$

bound. The robustness for the traditional random network is worse than that of the LCR random network, with about 80 % communication pairs failed when the attack radius is 0.3. Another observation is that the lower bounds of the two models are very loose, which could be explained by the derivations of the lower bounds in Sect. 4.4.

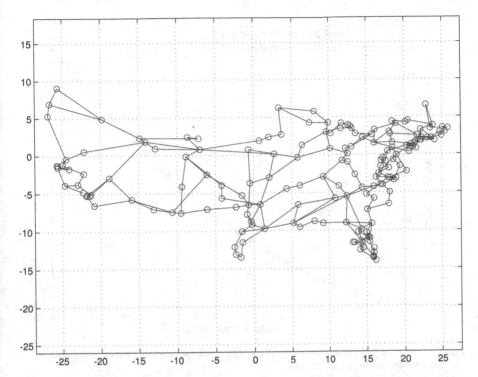

**Fig. 5.6** The fiber backbone of Level 3 Communications operated by a major US network provider

In Fig. 5.6, we show the fiber backbone of Level 3 Communications [56], which is a major US network service provider. There are totally 170 nodes connected by more than 240 links. In the following part of the simulation, we consider the area $[-30, 25] \times [-10, 15]$ as the network area, to avoid the invalid attack on the spare region. In Fig. 5.7, we present the simulation result of the expected number of attacked links of the fixed network case on this fiber network by assuming that the topology information is known. The result shows that the theoretical result derived in the fixed network case matches the simulation results on this fiber network.

In Fig. 5.8, we draw the expected number of attacked links in the fiber network of Level 3 Communications and compare it with the theoretical results by assuming that the network is a LCR random network. The parameters are set as $\lambda = 170, p = 0.137$, which are derived from the node number and the link number of this network. We assume that the communication range of the LCR random network is the length of the link that is longer than 90 % of the total number of the links in this fiber network. Thus we have that the communication range $l' = 4.6$ and the normalized communication range in the LCR random network model can be calculated as $l = l'/50 = 0.092$. We observe that the simulation result on this fiber network matches the theoretical result very well and the result is not sensitive to the changing of $l'$. However, in Fig. 5.9, the simulation result on the fiber network

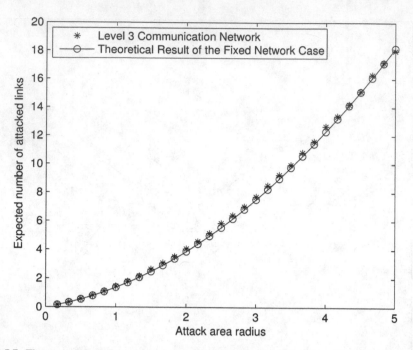

**Fig. 5.7** The expected number of attacked links in the fiber network of Level 3 Communications and the results of the fixed network case

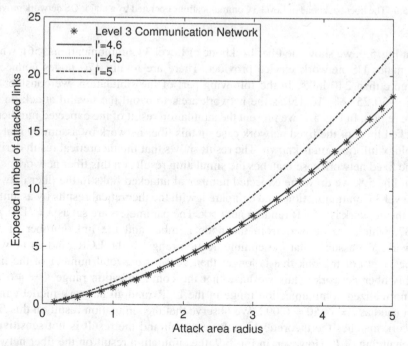

**Fig. 5.8** Expected number of attacked links in the fiber network of Level 3 Communications and the results of the LCR random network

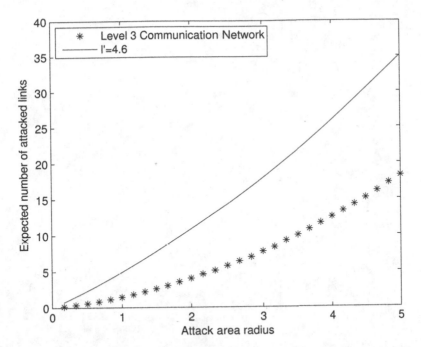

**Fig. 5.9** Expected number of attacked links in the fiber network of Level 3 Communication and the results of the traditional random network

does not match with the theoretical results under the assumption that the network is a traditional random network with the same parameters. Therefore, we conclude that the LCR random network model matches this fiber network, and without the exact knowledge of the node locations and the topology of a network, but only the high-level statistic information, such like the node number, the link number, and the normalized communication range, is enough to predict the expected number of attacked links in this real-world network to an area attack.

# Chapter 6
# Conclusion

In this monograph, we first started with an literature overview of the network reliability study for networks under various attacks and introduced the concept of area attacks, under which all of the nodes and links in a certain area are destroyed. For both the LCR and traditional random networks, we first studied the link-level robustness by deriving the average probability that an arbitrary link is attacked and the expected number of attacked links. We then studied the network-wide robustness under an area attack, where we gave four key properties that a good measure should have and accordingly proposed a new measure based on the percentage of surviving communication pairs. Simulation results showed that the LCR random network is more robust than the traditional random network under area attacks.

Q. Zhou et al., *Network Robustness under Large-Scale Attacks*, SpringerBriefs in Computer Science, DOI 10.1007/978-1-4614-4860-0_6, © The Authors 2013

# Appendix

In this Appendix, we derive $f_{RN}(x)$ and prove that the expected path length in the traditional random network is 0.5214. We assume that two nodes $n_1(A,C)$ and $n_2(B,D)$ are randomly located in $\mathcal{W}$. Thus, the length of link $n_1 n_2$ is $\sqrt{(A-B)^2+(C-D)^2}$, where $A$, $B$, $C$, and $D$ are independently and identically distributed (*i.i.d.*) with the following PDF

$$f_A(a) = \begin{cases} 1, & a \in [0,1] \\ 0, & \text{otherwise} \end{cases} \tag{A.1}$$

We define a new random variable $T$ as $A - B$, with the PDF given by

$$\begin{aligned} f_T(t) &= \int f_A(a)f_A(a-t)da \\ &= \int_0^1 1 \cdot f_a(a-t)da \\ &= \begin{cases} 1+t, & t \in [-1,0] \\ 1-t, & t \in [0,1] \end{cases} \end{aligned} \tag{A.2}$$

Define $Y$ as $Y = T^2 = (A-B)^2$. We have

$$\begin{aligned} P(Y \leqslant y) &= P(t^2 \leqslant y) \\ &= P(-\sqrt{y} \leqslant t \leqslant \sqrt{y}) \\ &= \int_{-\sqrt{y}}^{\sqrt{y}} f_T(t)dt \\ &= \int_0^{\sqrt{y}} f_T(t)dt + \int_{-\sqrt{y}}^0 f_T(t)dt \\ &= \int_0^{\sqrt{y}} 1-tdt + \int_{-\sqrt{y}}^0 1+tdt \end{aligned}$$

Q. Zhou et al., *Network Robustness under Large-Scale Attacks*, SpringerBriefs in Computer Science, DOI 10.1007/978-1-4614-4860-0, © The Authors 2013

$$= \left( t - \frac{t^2}{2} \right) \Big|_0^{\sqrt{y}} + \left( t + \frac{t^2}{2} \right) \Big|_{-\sqrt{y}}^0$$

$$= \sqrt{y} - \frac{y}{2} + \sqrt{y} - \frac{y}{2}$$

$$= 2\sqrt{y} - y. \tag{A.3}$$

Thus, the PDF of $Y$ is given by

$$f_Y(y) = (2\sqrt{y} - y)'$$

$$= y^{-\frac{1}{2}} - 1, \quad y \in [0,1]. \tag{A.4}$$

We denote $Z = (A - B)^2 + (C - D)^2$. From (A.4), the PDF of $Z$ can be expressed as

$$f_Z(z) = \int_{-\infty}^{+\infty} f_Y(y) f_Y(z - y) dy. \tag{A.5}$$

When $z \in [0,1]$,

$$f_Z(z) = \int_0^z (y^{-\frac{1}{2}} - 1)((z - y)^{-\frac{1}{2}} - 1) dy$$

$$= \pi - 4z^{\frac{1}{2}} + z. \tag{A.6}$$

When $z \in [1,2]$,

$$f_Z(z) = \int_{z-1}^1 (y^{-\frac{1}{2}} - 1)((z - y)^{-\frac{1}{2}} - 1) dy$$

$$= -2 \arcsin \left( \frac{-2 + z}{z} \right) - 2 + 4(z - 1)^{\frac{1}{2}} - z. \tag{A.7}$$

Let $X$ denote the length of the link $n_1 n_2$; thus $X = \sqrt{(A - B)^2 + (C - D)^2} = \sqrt{Z}$ and we have

$$P(X \leqslant x) = P(\sqrt{z} \leqslant x)$$

$$= P(0 \leqslant z \leqslant x^2)$$

$$= \int_0^{x^2} f_Z(z) dz. \tag{A.8}$$

When $x \in [0,1]$,

$$P(X \leqslant x) = \int_0^{x^2} \pi - 4z^{\frac{1}{2}} + z dz. \tag{A.9}$$

By taking the derivative of (A.9), we obtain the following PDF of the length for link $n_1n_2$ with $x \in [0,1]$:

$$f_{RN}(x) = f_X(x) = 2\pi x - 8x^2 + 2x^3. \tag{A.10}$$

When $x \in [1, \sqrt{2}]$,

$$P(X \leqslant x) = \int_0^1 \pi - 4z^{\frac{1}{2}} + z dz$$

$$+ \int_1^{t^2} -2\arcsin\left(\frac{-2+z}{z}\right) - 2 + 4(z-1)^{\frac{1}{2}} - z dz. \tag{A.11}$$

By taking the derivative of (A.11), we obtain the following PDF of the length for link $n_1n_2$ with $x \in [1, \sqrt{2}]$:

$$f_{RN}(x) = f_X(x) = 2x\left(-2\arcsin\left(\frac{-2+x^2}{x^2}\right) - 2 + 4(x^2-1)^{\frac{1}{2}} - x^2\right). \tag{A.12}$$

With numerical integration, the expectation of $X$ is obtained as 0.5214.

# References

1. Boesch FT, Harary F, Kabell JA (1981) Networks 11(1):57–63. doi:10.1002/net.3230110106
2. Phillips CA (1993) In: Proceedings of ACM symposium on theory of computing, vol 25. ACM, New York, p 776. doi:10.1145/167088.167286
3. Wood RK (1993) Math Comput Model 17(2):1. doi:10.1016/0895-7177(93)90236-R
4. Bingol L (2001) A lagrangian heuristic for solving a network interdiction problem. Master's thesis, Naval Postgraduate School
5. Royset JO, Wood RK (2008) INFORMS J Comput 19(2):175. doi:10.1287/ijoc.1060.0191
6. Motter AE, Lai YC (2002) Phys Rev E 66(6):065102. doi:10.1103/PhysRevE.66.065102
7. Sachtjen M, Carreras B, Lynch V (2000) Phys Rev E 61(5):4877. doi:10.1103/PhysRevE.61.4877
8. Goh KI, Kahng B, Kim D (2002) Phys Rev Lett 88(10):108701. doi:10.1103/PhysRevLett.88.108701
9. Watts D (2002) Proc Natl Acad Sci 99(9), 5766. doi:10.1073/pnas.082090499
10. Moreno Y, Gomez J, Pacheco A (2002) Europhys Lett 58(4):630. doi:10.1209/epl/i2002-00442-2
11. Willinger W, Govindan R, Jamin S, Paxson V, Shenker S (2002) Proc Natl Acad Sci 99(1), 2573. doi:10.1073/pnas.012583099
12. Callaway DS, Newman MEJ, Strogatz SH, Watts DJ (2000) Phys Rev Lett 85:5468. doi:10.1103/PhysRevLett.85.5468
13. Ball F, Mollison D, Scalia-Tomba G (1997) Ann Appl Probab 7(1):46. doi:10.1214/aoap/1034625252
14. Newman MEJ, Watts DJ (1999) Phys Rev E 60(6):7332. doi:10.1103/PhysRevE.60.7332
15. Cohen R, Erez K, ben Avraham D, Havlin S (2000) Phys Rev Lett 85:4624. doi:10.1103/PhysRevLett.85.4626
16. Zhou Q, Gao L, Cui S (2010) In: Proceedings of IEEE GLOBECOM. IEEE, Piscataway, pp 368–372. doi:10.1109/GLOCOMW.2010.5700343
17. Foster JS, Gjelde E, Graham WR, Hermann RJ, Kluepfel HM, Lawson RL, Soper GK, Wood LL, Woodard JB (2004) Report of the commission to assess the threat to the united states from electromagnetic pulse (emp) attack. http://www.empcommission.org/docs/empc_exec_rpt.pdf. Accessed 20 Apr 2012
18. Radasky W (2007) High-altitude electromagnetic pulse (hemp): a threat to our way of life. http://www.todaysengineer.org/2007/Sep/HEMP.asp. Accessed 20 Apr 2012
19. Ozdamar L, Ekinci E, Kucukyazici B (2004) Ann Oper Res 129:217. doi:10.1023/B:ANOR.0000030690.27939.39
20. Radwan AE, Hobeika AG, Sivasailam D (1985) Inst Transp Eng J 55(9):25
21. Young S, Balluz L, Malilay J (2004) Sci Total Environ 322:3. doi:10.1016/S0048-9697(03)00446-7

22. Feng QS, Chen JF, Ai MY (2010) J Acta Petrolei Sinica 1:139
23. Wang JW, Rong LL (2009) Elsevier Saf Sci 47(10):1332. doi:10.1016/j.ssci.2009.02.002
24. Pastor-Satorras R, Vazquez A, Vespignani A (2001) Phys Rev Lett 87(25):258701. doi:10.1103/PhysRevLett.87.258701
25. Cohen R, Erez K, Ben-Avraham D, Havlin S (2001) Phys Rev Lett 86(16):3682. doi:10.1103/PhysRevLett.86.3682
26. Gligor VD (1984) Proc IEEE Trans Softw Eng **10**(3), 320. doi:10.1109/TSE.1984.5010241
27. Peng T, Leckie C, Ramamohanarao K (2007) ACM Comput Surv 39(1):1. doi:10.1145/1216370.1216373
28. Mirkovic J, Reiher P (2004) ACM Comput Commun Rev 34(2):39. doi:10.1145/997150.997156
29. Burch H, Cheswick B (2000) In: Proceedings of the USENIX LISA. USENIX Association, Berkeley, pp 319–327
30. Savage S, Wetherall D, Karlin A, Anderson T (2000) ACM Comput Commun Rev 30(4):295. doi:10.1145/347057.347560
31. Snoeren AC, Partridge C, Sanchez LA, Jones CE, Tehakountio F, Kent ST, Strayer WT (2001) ACM Comput Commun Rev 31(4):3. doi:10.1145/964723.383060
32. Albert R, Jeong H, Barabasi AL (1999) Nature 6749:130
33. Barabasi AL, Albert R (1999) Science 286(5439):509
34. Barabasi AL, Albert R, Jeong H (2000) Phys A Stat Mech Appl 281(1–4):69. doi:10.1016/S0378-4371(00)00018-2
35. Faloutsos M, Faloutsos P, Faloutsos C (1999) Comput Commun Rev 29(4):251. doi:10.1145/316194.316229
36. Li W, Cai X (2004) Phys Rev E 69(4):046106. doi:10.1103/PhysRevE.69.046106
37. Albert R, Jeong H, Barabasi AL (2000) Nature 406(6794):378. doi:10.1038/35019019
38. Neumayer S, Zussman G, Cohen R, Modiano E (2009) In: Proceedings of IEEE INFOCOM. IEEE, Piscataway, pp 1566–1574. doi:10.1109/INFCOM.2009.5062074
39. Brown G, Carlyle WM, Salmeron J, Wood K (2005) Tutorials in operations research. INFORMS, Hanover, pp 102–123
40. Brown G, Carlyle M, Salmeron J, Wood K (2006) Informs 36(6):530. doi:10.1287/inte.1060.0252
41. Kephart J, White S (1991) IEEE computer society symposium on research in security and privacy. IEEE, Los Alamitos, pp 343–359. doi:10.1109/RISP.1991.130801
42. Gallos LK, Cohen R, Argyrakis P, Bunde A, Havlin S (2005) Phys Rev Lett 94(18):188701. doi:10.1103/PhysRevLett.94.188701
43. Magoni D (2003) IEEE J Sel Areas Commun 21(6):949. doi:10.1109/JSAC.2003.814364
44. Agarwal P, Efrat A, Ganjugunte S, Sankararaman S, Hay D, Zussman G (2011) In: Proceedings of IEEE INFOCOM. IEEE, New York, pp 1521–1529. doi:10.1109/INFCOM.2011.5934942
45. Sen A, Banerjee S, Ghosh P, Shirazipourazad S (2009) In: Proceedings 47th annual allerton conference on communication, control, and computing. IEEE, Piscataway, pp 1430–1437. doi:10.1109/ALLERTON.2009.5394506
46. Liu J, Jiang X, Nishiyama H, Kato N (2011) IEEE Trans Veh Technol 60(5):2253. doi:10.1109/TVT.2011.2114684
47. Kim K, Venkatasabramanian N (2010) In: Proceedings of IEEE GLOBECOM. IEEE, Piscataway, pp 1–5. doi:10.1109/GLOCOM.2010.5685229
48. Neumayer S, Modiano E (2010) In: Proceedings of IEEE INFOCOM. IEEE, New York, pp 1–9. doi:10.1109/INFCOM.2010.5461984
49. Dekker AH, Colbert BD (2004) Australas Comput Science Conf 26:359
50. Mathew P (2003) Random geometric graphs. Oxford University Press, Oxford
51. Schlick T (2002) Molecular modeling and simulation: an interdisciplinary guide. Springer, New York
52. Manna SS, Sen P (2002) Phys Rev E 66(3):066114. doi:10.1103/PhysRevE.66.066114
53. Ebel H, Mielsch L, Bornholdt S (2002) Phys Rev E 66:35103. doi:10.1103/PhysRevE.66.035103

54. Meester R, Roy R (1996) Continuum percolation. Cambridge University Press, New York
55. Chung F, Lu L (2003) Internet Math 1(1):91. doi:10.1080/15427951.2004.10129081
56. Level 3 Communication Corp (2012) Level 3 communications network map. http://www.level3.com/en/resource-library/maps/level-3-network-map/. Accessed 20 Apr 2012